YOUR KNOWLEDGE HAS VALUE

Bibliographic information published by the German National Library:

The German National Library lists this publication in the National Bibliography; detailed bibliographic data are available on the Internet at http://dnb.dnb.de .

Imprint:

Copyright © 2016 GRIN Verlag, Open Publishing GmbH
Print and binding: Books on Demand GmbH, Norderstedt Germany
ISBN: 9783668379633

Madhusudhan Pandey

Single phase sinusoidal input to non-sinusoidal output Cycloconverter

GRIN Publishing

GRIN - Your knowledge has value

Since its foundation in 1998, GRIN has specialized in publishing academic texts by students, college teachers and other academics as e-book and printed book. The website www.grin.com is an ideal platform for presenting term papers, final papers, scientific essays, dissertations and specialist books.

Visit us on the internet:

http://www.grin.com/

http://www.facebook.com/grincom

http://www.twitter.com/grin_com

Single phase sinusoidal input to non-sinusoidal output Cycloconverter with output frequency, $f_o = \frac{f_{in}}{n}$, where $n = 2,4,6,8\ldots\ldots$ using reduced number of Thyristor in P and N Converters.

Madhusudhan Pandey

TU, IOE, Pashchimanchal Campus

Abstract: A simple single phase step down cycloconverter with reduce numbers of Thyristors in P and N converter is described. A shunt capacitor connected with load is used to convert sinusoidal input at one frequency to non-sinusoidal output at another frequency.

Content

Section I: Introduction

Cycloconveters are the arrangement of power electronics which is used to convert Alternating Current (AC) at one frequency to AC at another frequency. The power circuit diagram of single phase bridge type cycloconverter is the basic circuit diagram to explain concept of cycloconverters. Fig. 1 shows the power circuit diagram [1][2]. There are two, P and N, converters for carrying positive and negative half cycle output signals.

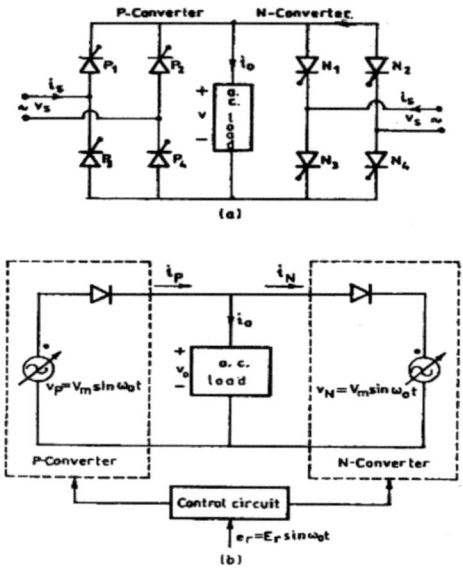

Fig. 1 Power circuit diagram of single phase step down bridge type cycloconverter.

Section II: Proposed diagram with reduce number of Thyristors

Fig. 2 shows circuit diagram of modified circuit diagram of Fig. 1. In this circuit diagram there are two thyristor T1 and T2 for P converter and T3 and T4 for N converter. For the sake of clearness in operation, operating principle of modified diagram can be explained more specifically.

2

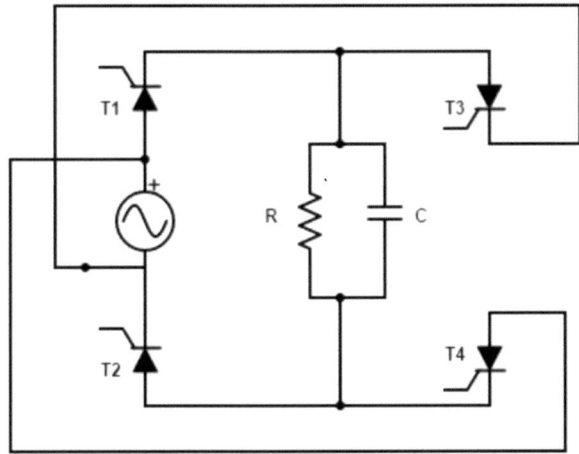

Fig. 2 Proposed Diagram of Modified
Cycloconverter with a Shunt Capacitor

A shunt capacitor is connected to resistive load (R). Suppose we want $220V(RMS), 50Hz$ (time period= $20ms$) to $25Hz$ (time period= $40ms$) cycloconverter. At first, T1 and T2 are ON at $0ms$, which opens valve of T1 and T2 for the flow of the first half cycle of AC input. Upto $5ms$ capacitor is charged with maximum input voltage ($V_m = 311.11V$). Now for $15ms$ (i.e remaining time period of first half of output waveform) capacitor discharges with suitable time for discharging (further clarification is done in Section III). After $20ms$ T3 and T4 are ON for negative half cycle of output. Same process of charging and discharging of capacitor takes place for negative cycle of output. Fig. 3 shows output waveform for this specified case.

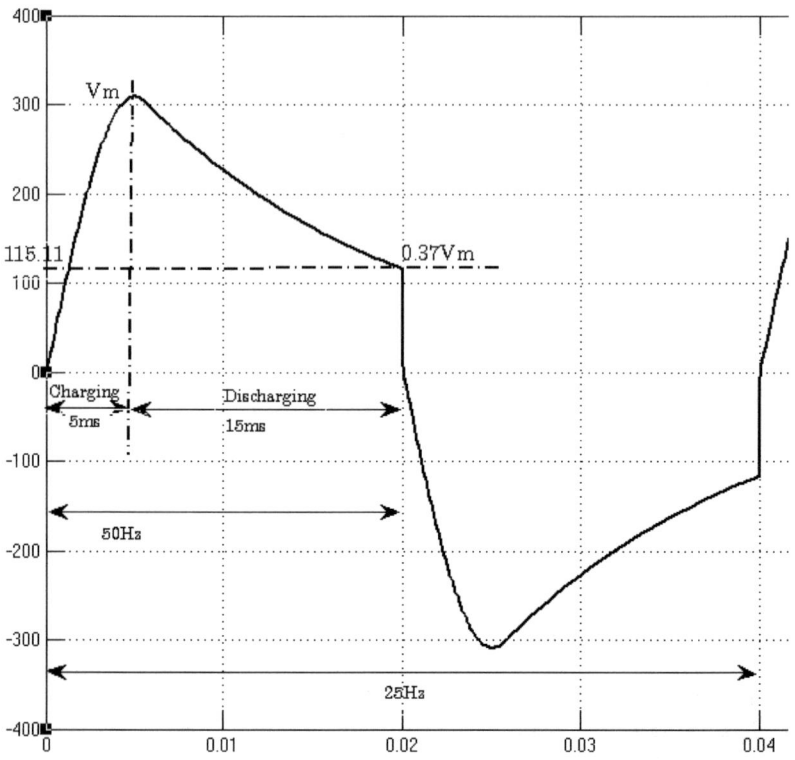

Fig. 3 Output of 50Hz to 25Hz Cycloconverter

Section III: Theory

A shunt capacitor is connected with R-load. In first half cycle of output, when Thyristors T1 and T2 are ON at 0ms, capacitor is charged upto time period $\frac{T_{in}}{4}$ ($or\frac{T_o}{4n}$). Now, we choose value of capacitor such that it discharges thorough load resistor for $\frac{3T_{in}}{4}$ ($\frac{3T_0}{4n}$), until voltage across capacitor (V_c) becomes $x\% ofV_m$ just before zero crossing of first half cycle of output voltage. We can also select a capacitor based on the value of V_c when it becomes say $37\% ofV_m$ (i.e. V_c at τ_D, discharging time constant of capacitor) at zero crossing of output voltage.

4

The voltage across capacitor when it discharges is given by [3]:

Or, $V_c = V_m e^{-(\frac{t}{\tau_D})}$

We know capacitor discharging time, $\tau_D = RC$

Or, $x\% of V_m = V_m e^{-(\frac{t}{RC})}$

After mathematical reduction we get as,

Or, $C = \dfrac{t}{R \ln(\dfrac{100}{x})}$ (1), where $t = \dfrac{T_{in}}{4}$ (or $\dfrac{T_o}{4n}$)

During negative half cycle when Thyristors T3 and T4 are ON at T_{in} (or $\dfrac{T_o}{2}$), the charging and discharging of capacitor continues but with reversal appearance than that of positive half cycle. In this way desired frequency of output voltage with non-sinusoidal waveform is obtained.

Specified Case:

Let us take an instance that input frequency to be $50Hz$ for rms input voltage of $220V$ where maximum input voltage $V_m = 311.11V$ and resistive load is 1Ω. A cycloconverter with output frequency $25Hz$, $12.5Hz$, $6.25Hz$, etc. can be obtain by choosing a suitable capacitor connected in shunt with the resistive load.

1. For $50Hz$ to $25Hz$, $T_{in} = 20ms$ and $T_0 = 40ms$. When T1 and T2 are ON at $0ms$, capacitor is charged upto $5ms$ with maximum voltage $V_m = 311.11V$. Now we have to select a capacitor that discharges upto $15ms$. Let, output voltage value just before zero crossing of first half cycle of output waveform (i.e. at $\dfrac{T_0}{2} = 20ms$) be $37\% of V_m$, which is output voltage at discharging time of capacitor, τ_D. Now, value of capacitor can be calculated as

Or, $\tau_D = RC$ (2)

As, $R = 1\Omega$ and $\tau_D = 15ms$

Equation (2) gives,

Or, $C = \tau_D * 1 = 15mF$.

5

Simulation, with above input parameters, is done in Simulink Environment of MATLAB and output voltage waveform is shown in Fig. 4.

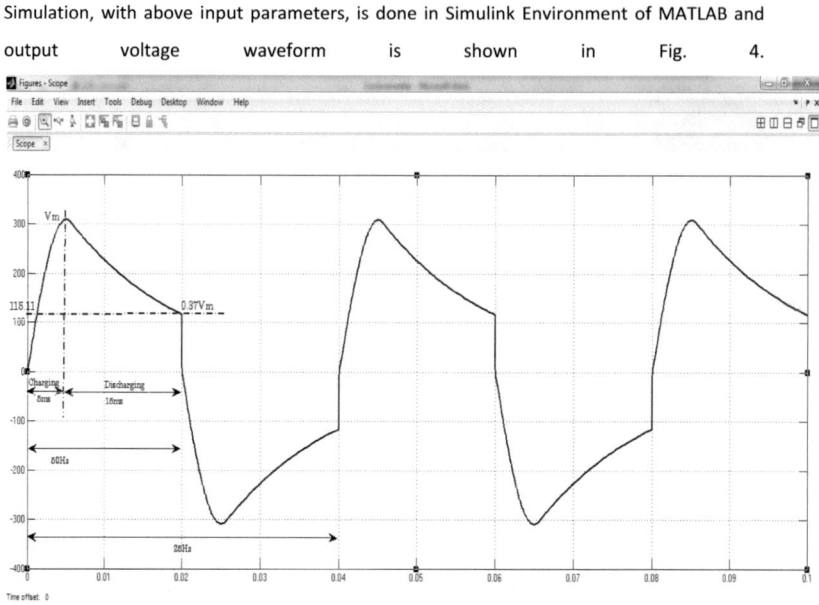

Fig. 5 Block Diagram of Modified Single Phase
Cycloconverter

Similarly, if just before zero crossing of output voltage we want output voltage across capacitor to be $x\% of V_m$ then for different values of x, values of capacitor are calculated by using equation(1) and shown in Table I.

Table: I

$x(in\%)$	$C = \dfrac{t}{R \ln(\dfrac{100}{x})}(inF)$
1. 10	0.0065
2. 20	0.0093
3. 40	0.0164
4. 60	0.0294
5. 70	0.0421
6. 80	0.0672
7. 90	0.1424

6

For several values of x, simulation result shows various output voltage waveform:

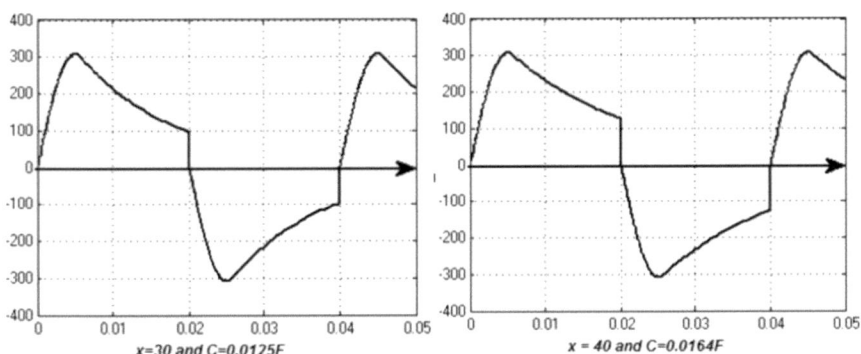

x=30 and C=0.0125F   x = 40 and C=0.0164F

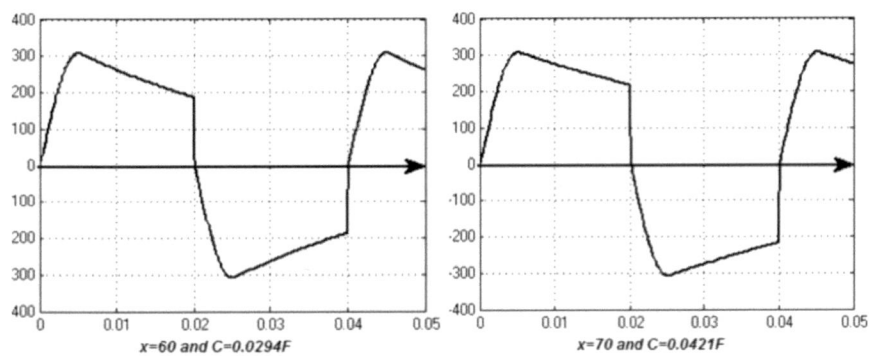

x=60 and C=0.0294F   x=70 and C=0.0421F

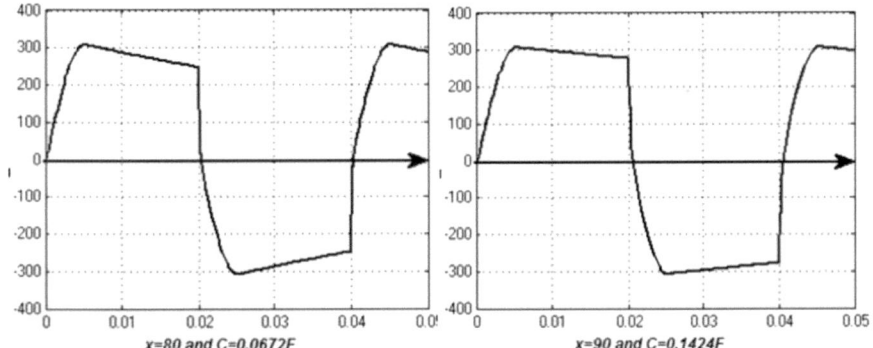

x=80 and C=0.0672F   x=90 and C=0.1424F

2. For $50Hz$ to $12.5Hz$, $T_{in} = 20ms$ and $T_o = 80ms$. In this case capacitor charges upto $5ms$ and discharge for $35ms$. Setting all parameter same, different values of capacitance of shunt capacitor can be calculated for different values of x as shown in Table II.

Table II

$x(in\%)$	$C = \dfrac{t}{R\ln(\dfrac{100}{x})}(inF)$
1. 10	0.0152
2. 40	0.0382
3. 70	0.0981
4. 90	0.3322

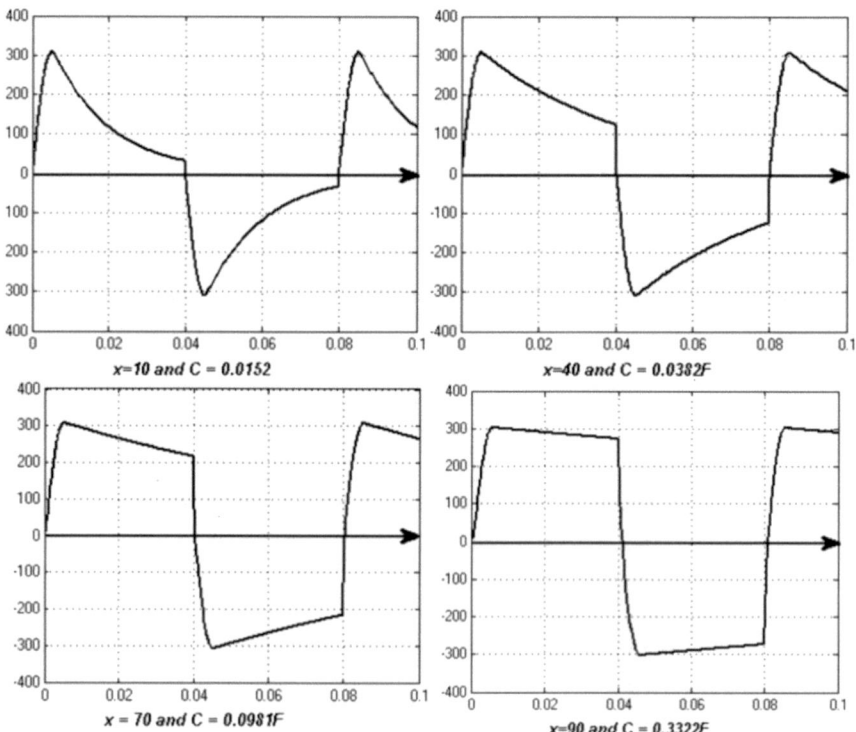

Section IV: Simulation in MATLAB Simulink:

Fig. 5 shows simulation block diagram made in MATLAB.

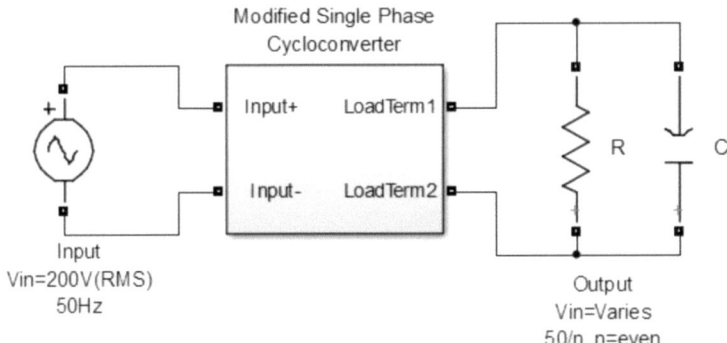

Fig. 5 Block Diagram of Modified Single Phase
Cycloconverter

The block diagram Modified Single Phase Cycloconverter in figure is a subsystem created using P and N converter which is shown in Fig. 6

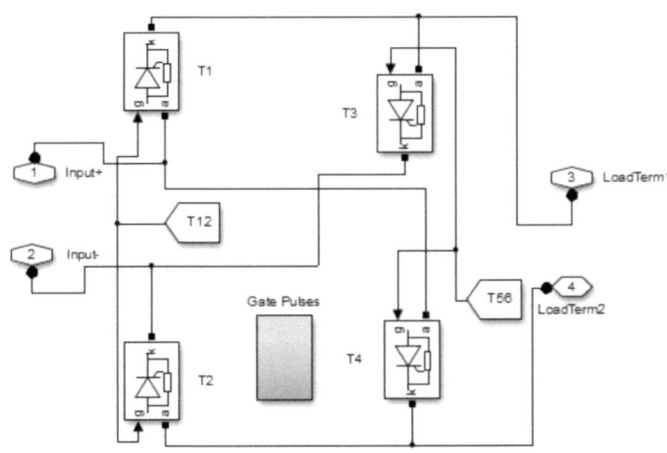

Fig. 6 P and N conveter of Modified Single Phase
Cycloconverter

Simulation file for 50Hz to 25Hz, x=60 and C=0.0294, can be downloaded from link below:

9

Various parameter like input voltage, value of capacitor C and gate pulses for thyristors can be changed to get the desired output frequency.

Section V: Conclusion

By reducing a number of thyristor in both P and N converters and choosing a suitable shunt capacitor at load terminal, a modified step down cycloconveter converts sinusoidal input at one frequency(f_{in}) to non-sinusoidal output another frequency (f_o), with $f_o = \dfrac{f_{in}}{n}$, with $n = 2,4,6\ldots\ldots$

Section VI: References

[1] Rashid, Muhammad H. *Power Electronics Handbook*. San Diego: Academic, 2001. Print.

[2]Pelly, B. R. *Thyristor Phase-controlled Converters and Cycloconverters: Operation, Control, and Performance*. New York: Wiley-Interscience, 1971. Print

[3] "Charging and Discharging a Capacitor." Charging and Discharging a Capacitor. N.p., n.d. Web. 10 Feb.2016.<http://macao.communications.museum/eng/exhibition/secondfloor/MoreInfo/2_3_5_ChargingCapacitor.html>.